Painting

Deri Robins

QEB Publishing

Copyright © QEB Publishing, Inc 2004

Published in the United States by
QEB Publishing, Inc.
23062 La Cadena Drive
Laguna Hills
Irvine
CA 92653

Library of Congress Control Number:
2004101532

ISBN 1-59566-046-1

Written by Deri Robins
Designed by Wladek Szechter/Jacqueline
Palmer
Edited by Sian Morgan/Matthew Harvey
Artwork by Melanie Grimshaw
Photographer Michael Wicks
With thanks to Nicola

Creative Director Louise Morley
Editorial Manager Jean Coppendale

Picture credits

The Art Archive /18 top Eileen Tweedy, Tate
Gallery, London/
Corbis /29 bottom Roger De La Harpe, Gallo
Images

Printed and bound in China

The words in **bold** are
explained in the Glossary
on page 30.

Contents

colored
pencils

Tools and paints

You only need a few brushes and a simple paint set to get started—but it's a good idea to build up a collection of some of the equipment below if you want to experiment with different effects.

Pencils

Soft-lead pencils (such as 9B) are useful for sketching the outlines of your pictures before adding paint. Use colored pencils to add detail (see page 10–11).

pencils

Brushes

Different brushes give different results. You need a fine, pointed one for details, a middle thickness one, and a really thick one for large areas. Nylon bristles are best for thin paint, and hogs' hair brushes are better for thick paint. Use an old toothbrush for spattering (see pages 28–29), and sponges and rags for dabbing (see pages 12–13).

Palettes and pots

You can buy artists' **palettes** for mixing colors—but an old plastic tray, an old plate, or piece of smooth wood are just as good. Use jars full of water to clean your brushes.

Bits and pieces

Use newspaper to protect your work counter and paper towels to dry your brushes. Collect or draw ideas in a notebook or **sketchbook** when you are away from home. Paste things that inspire you, such as colored paper, leaves, and magazine cuttings.

brushes

palette

watercolors

acrylic paints

ink

gouache paints

Paint

Poster paints are ideal for big, bold paintings, or to mix with other things to create texture (see pages 22–23). Use them straight from the container or thin them down with water.

Watercolors are delicate colors that make great landscapes. They come in tubes or blocks, are easy to carry and handy for color sketches.

Gouache paints produce strong, vibrant colors that are good to use on colored papers.

Paper

It's important to use good paper. Smooth sketch paper is best. The colors don't sink into the surface and become dull. Thick paint on rough paper can be interesting. Watercolor paper is thick, so that it doesn't wrinkle when it gets wet.

art paper

poster paints

Color mixing

Mixing colors is one of the most important things you will need to do as a painter. Take time to read these pages, and try out lots of color experiments of your own.

Make it up

Nearly every color you can think of is made up from just three **primary** colors: red, blue, and yellow. Here's how to mix paints to get all the colors you need.

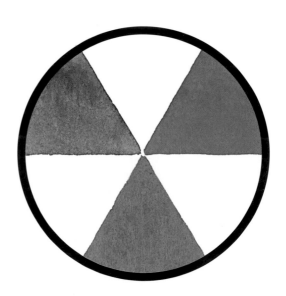

Primary colors

Red, yellow, and blue are called the primary colors because you can't make them from any other colors.

Secondary colors

Mix any two primary colors together and you get a **secondary** color. They are orange, green, and purple.

Color wheel

The inner rim of the wheel shows what happens when you mix primary colors: red and yellow make orange; red and blue make purple; and yellow and blue make green.

The outer rim shows what happens when you mix secondary colors together. You get six more colors.

Make your own version of this color wheel and try different color combinations.

TIP

Make your own color charts, like the ones at hardware stores, then personalize them. Draw small rectangles on a sheet of paper. Fill each with a different **shade**—choose a range of blues or a range of pinks, for example. Give them a name, such as "princess pink," or "sea-haze green." Remember to make a note of the colors you used to mix them—and how much of each color you used.

Complementary colors

Colors that are opposite each other on the color wheel are called **complementary** colors. Put them together and they can make paintings exciting and make the colors seem deeper.

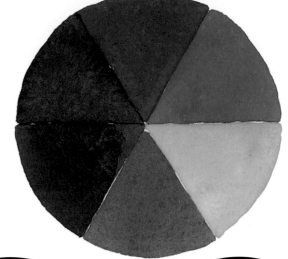

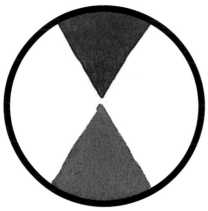

green and red

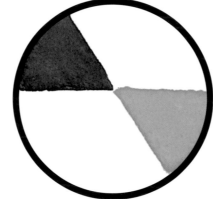

yellow and violet

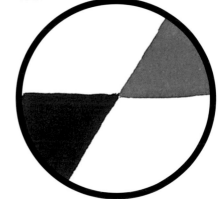

blue and orange

Making colors darker

Don't just add black to make a darker color: often this just makes it look dull and boring. Instead, add a little bit of a different, darker color. Experiment on scrap paper until you get a shade you really like.

Lighten up

Add more water to watercolor paint for lighter **shades**. With poster or powder paint, just add a little white paint at a time, until you get the shade you want.

Warm and cool

Colors can be described as "cool" (such as green and blue) or "warm" (orange, yellow, and red). You can use these colors to create cool, calm paintings or hot, exciting ones.

Grays and browns

Black and white is only one way to make gray. You get much more interesting shades of gray by mixing various amounts of the three primary colors together or two of the complementary colors.

There are many different shades of brown, too. See how many shades you can come up with.

Paint effects

Different brush strokes can give you very different effects. Experiment to find which you enjoy using, and which brushes are best for different types of painting.

Look at the way the artist has painted these animals and see how many different effects you can get using different brush strokes, colors, and materials.

This cat was painted quickly and in a loose style so that the colors blended together while still wet—it looks very effective from a distance.

The tip of a thin brush was used to paint the fine feathers on this toucan.

When painting this parrot, the artist first colored it in watery paint. The details—beak, eyes, and feathers—were added using drier paint and thin brushes.

Highlights

White highlights make these glasses and balloons look shiny. You can get this effect by leaving that part of the picture white (without paint), or by adding a few strokes of white paint when it is dry.

Different textures

You don't have to paint in solid color. You can use thin brushes to do lots of different-colored dots, or square-ended brushes to make lines of color.

Try using a dry, stubby brush and dab nearly dry paint over different areas of your picture to make interesting textures, such as on this gingerbread man.

TIP

Keep experimenting with different paint effects to use in your pictures. You could try using a dry brush with nearly dry paint to make interesting textures.

No-brush art

You don't need a brush to paint! What about using sponges, rags, scraps of paper and fabric, and ordinary objects you can find around the house? You can even use your hands and fingers.

Home-made brushes

It's easy to make your own brushes. Try the examples shown below, or make up your own. What do the effects make you think of? How could you use them in your paintings?

TIP

Try different effects on pieces of scrap paper. Label your results so you can remember how you did them. Experiment with different paints and materials. Use your fingers and hands, too!

Cotton wool makes puffs of steam or clouds.

Drag a twig through wet paint for a rough grass effect.

Scrunched-up paper towels makes interesting textures and creases.

A sponge is good for making grass or soft animal fur.

Dragging paint

You can drag combs, plastic knives and forks, or the edge of a piece of cardboard through the paint to make patterns and markings.

Fingertip rabbit

1 Lightly draw the outline of the rabbit in pencil.

2 Now fill in the outline with paint, using your fingers. Let each color dry before adding the next one.

The grass in this picture was made with a mixture of thumb prints and a twig.

From a distance

Ever since the first artist picked up a brush, people have tried to paint the natural world around them. Here is a great project that will help you capture lots of different views in your paintings.

Paint a 3-D theater

This **3-D** theater is made up of cards, one behind the other, to make the scenery look convincing. Just as in a painting, you should make the front scenery strong using bold colors and the faraway parts paler.

1 Cut two squares of cardboard, each 7 inches square. Use a ruler to divide them into ten ¾ inch lines.

2 Fold along the lines to make two **accordian** shapes for the sides of your theater.

3 Glue the accordian sides onto a cardboard base. Cut out five rectangles, each 7 x 9 inches. Paint one blue for the sky. Glue it to the sides to make the **backdrop**. Cut another rectangle into a frame shape and glue it to the front.

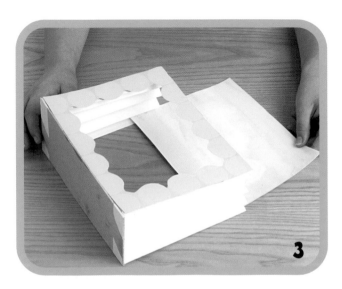

2

3

4 The three remaining rectangles are for your scenery. Draw a country scene on one softly in pencil, with a house at the top. The other two pieces should be rolling hedges. Draw these so that they are not as tall as the house scenery. The smaller scenes will slot in the front.

5 Now color the scenery. Remember to use soft, pale colors in the distance, and strong, sharp details at the front. Finally, cut the pieces out, ready to slot into place in your 3-D theater.

Create different sets of scenery for your theater. What about a city at night?

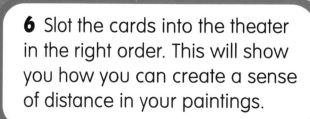

6 Slot the cards into the theater in the right order. This will show you how you can create a sense of distance in your paintings.

Painting with dots

Do you know that there is a way to "mix" colors so that you don't have to blend the paint on a **palette**? Put dots of different colors next to one another, and when you stand back, the colors appear to merge together!

Dots and dashes

Practice using dots and dashes of color before you begin painting.

1 Start making strokes with your first color, leaving spaces between the strokes.

2 When the first color is dry, fill in the spaces with strokes of a second color. Now stand back from the image and look at it from a distance—what effect have you achieved? What happens if you add more dots to the image?

You can use dots to make colors merge.

16

Make a colorful dot painting

WHAT YOU NEED
- Paper
- A pencil
- Poster paints
- Thin brushes

1 Use a soft pencil to sketch this picture onto paper.

2 Color each piece of fruit with dots of color. Just use one color for each piece of fruit at first. Make the apple red, the banana yellow, and the pear green.

3 Now create **shade** and highlights by adding dots of a darker color. You can also create shade by adding more dots of the first color.

4 When the fruit is complete, add a background to your picture in a warm, gentle color.

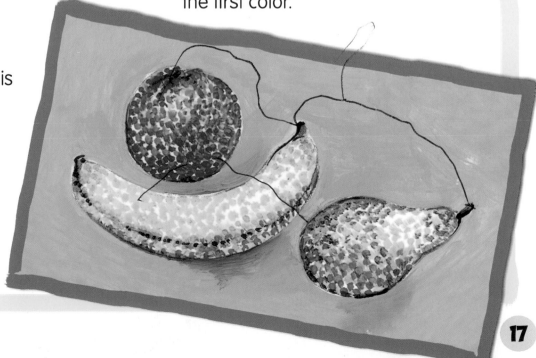

17

Abstract art

You don't have to make your paintings look realistic if you don't want to. Many famous artists painted what they saw using flat patterns, bright colors, and interesting shapes. This is called **abstract** art.

Different styles

Many artists moved away from realistic painting and developed styles of their own. **Pablo Picasso** often used a style called **cubism**, where everything is made of squares. **Henri Matisse** made images with blocks of color.

**L'Escargot (The Snail)
by Henri Matisse**

TIP

Henri Matisse often used torn and cut paper to make bold **collage** images.
Try making a collage of your artwork, using pieces of colored paper instead of paint.

Make an abstract painting

Decide what you are going to paint—for example, the view from your window, a scene from a postcard, a person, or an animal.

1 Lightly sketch the main parts of the picture with a pencil. Just draw the houses and trees as simple shapes. In this painting, the artist hasn't tried to put anything in the right place—he has just used them to make a pattern of shapes and colors.

2 Paint the finished picture, using bold colors. Apply the paint in bold blocks of color, without **shading**.

Painting patterns

Flowers, berries, and leaves make great patterns. Look for examples in your yard or local park. Look through magazines, wallpaper samples, and fabric remnants to see how artists use natural forms in their designs. You can also make your own natural patterns.

WHAT YOU NEED
- Pencil
- Thick paper
- Poster paints
- Scissors
- Glue

Design your own wallpaper or material

1 Look through your collection of leaves, flowers, and plants and sketch some of your favorite shapes.

2 Put the shapes together to make a pattern—cut them up and arrange them on the page, glue them in place.

3 For a repeat design, draw a grid with a pencil and ruler. This will help you to keep the shapes straight, the same size, and in the right place. Your shapes can be realistic or imaginative.

4 Paint the pattern using poster paint. Which works best—two or three colors, or more? (Look at the tips on using color on page 8–9.) Why not try some different designs?

TI

English **Victorian** artists often used nature in their designs. **William Morris** used plants, flowers, and birds in wallpaper, **tapestries**, stained-glass windows, tiles, and embroidery.

Dragging and combing

WHAT YOU NEED
- Thick paint
- A wide brush
- Posterboard
- Pieces of cardboard or plastic packaging
- Scissors • Glue

Dragging and combing pictures and designs in thick, wet paint is a fun way to create images.

1 Cut comb shapes out of cardboard. Make some with thick teeth, and some with small teeth.

2 Brush a thick coat of paint onto the paper in different colors. Divide the page into roughly equal bands of color.

3 While the paint is still wet, drag your card combs through the paint to make waves and lines. The colors will carry over into the other bands, and you will see the paper beneath.

Scraper art

As well as making patterns, you can use your scrapers to make a picture.

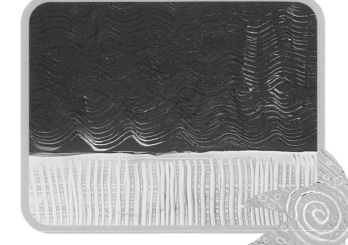

1 Paint a wide band of yellow across the bottom of the paper and fill in the rest of the page with blue paint. While the paint is still wet, use a cardboard comb to make swirly waves in the sea. Use a comb to make **horizontal** lines across the sand, then comb upward for a **checked** effect.

2 On a separate piece of paper, use paint and combs to make sea animals and plants.

3 When these are dry, cut them out and glue them to the sea and sand background.

Can you think of any other swirly pictures to make? What else could you use to make marks in the paint?

23

Watery painting

You can get some wonderful effects if you brush watery paint onto wet paper. The colors blend together, giving a very soft result—perfect for painting skies and seas.

WHAT YOU NEED
- Watercolor paint
- Soft wide brush
- Thick paper
- Water in jar

Two-color painting

Practice using two colors to make watery paintings.

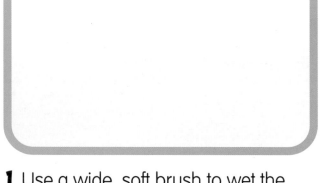

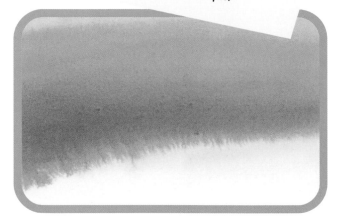

1 Use a wide, soft brush to wet the paper. Then paint a wide strip of yellow across the bottom half of the paper.

2 Before it dries, dip your brush in the second color. Brush this gently over the top so that the two colors merge softly. Work quickly before the paint dries!

TIP

If you don't want your paintings to go crinkly when they dry, stretch the paper first. Dip the paper very quickly in water, and tape it to a board or piece of cardboard. Once dry, it is ready to use.

Paint a sunset

1 Brush water all over the paper with a wide, soft brush. Paint the background in yellow and, while the paint is wet, add some orange streaks.

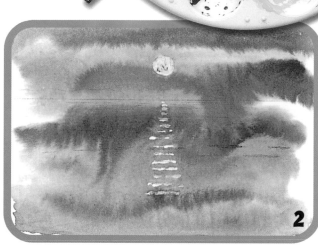

2 Now use some dark blue paint to add a band at the top of the image. Let this dry before using white paint to add the sun and its reflection in the water.

3 Use black paint to make the strip of land about three-quarters of the way up the picture. Add some ripples in light blue in the **foreground**.

4 When the paint is dry, use a small brush and dry black paint to add the boats. They appear as **silhouettes** in the sunset, with dark reflections in the water.

Spatter art

Some artists don't use their brushes to paint objects, they just throw the paint onto their pictures! It is a great way of making a fun, lively image with lots of color.

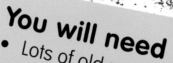

Make your own messy masterpiece

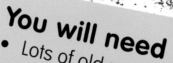

You will need
- Lots of old newspaper
- A piece of paper
- Paint
- Brushes
- An old toothbrush
- Cardboard

1 Cover the floor with newspaper and wear an apron or artist's smock. Put your sheet of paper in the middle.

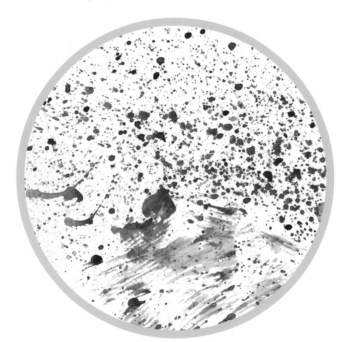

2 Flick runny paint onto the paper with brushes. Dip the toothbrush into paint, and run the cardboard over the bristles while you point it at the paper.

3 Add blobs, drizzles, and splats of different color, until you like the result. Leave some white background showing through.

Spatter stencils

Spattering and stencils make sharp pictures on messy backgrounds.

1 Cut some shapes out of cardboard, realistic or fantasy ones. Arrange them on a piece of paper.

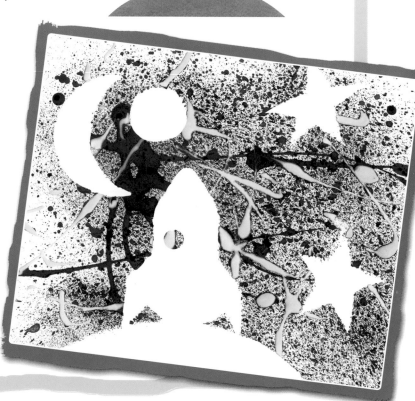

2 Spatter the paint over the paper as before. Carefully lift up the shapes when you have finished.

TIP

If you want to make one or more areas of your picture darker, dip a toothbrush in paint then hold it near the area you want to make darker and keep flicking the paint at the paper using your thumb.

Going BIG!

Choose strong, simple images for a large painting or mural. The trick is to plan them first of all on a sheet of paper. Then it's easy to scale them up in size.

What you need
- A4 paper
- A large piece of paper
- Ruler • Pencil
- Paint • Brushes

Using a grid to scale up

1 Sketch out your drawing or design on a standard sheet of paper. Paint in the colors you want to use.

2 When the paint has dried, draw a grid over the painting. Use a ruler and pencil to divide the picture into 2-inch squares along the top and one side. Join the marks to make squares all over. If the picture doesn't divide exactly into 2-inch squares, make the bottom rows slightly smaller.

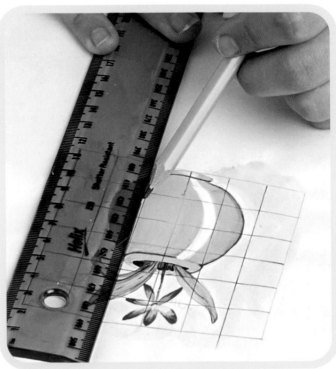

3 Now take your big piece of paper, and divide it into a grid with the same number of squares—but this time make the squares much bigger. Copy your picture, square by square, in pencil. When you are happy with it, paint it. Use your small painting as a guide to get the colors right.

TIP

If you don't want to spoil your first painting, tape a piece of tracing paper over the top, and draw the grid on this.

TIP

You can use this technique to create really huge pictures. Lots of different cultures make large images to decorate the walls of their houses or streets. Try making a large image for your classroom wall.

Glossary

3-D stands for three-dimensional—when an object has or appears to have height, width, and depth

abstract pictures that do not look like real objects

accordian fold folding a piece of paper or card into a small folds so that it has a zigzag shape

backdrop the scenery in a theater at the back of the stage

checked having a pattern made up of squares

collage making pictures or patterns using different materials, such as paper and cloth, which are glued onto a background

complementary colors colors that work well together

cubism a form of art in which pictures were made up of fragments of images

foreground the area at the front of a picture

horizontal running across the page, rather than up and down

Henri Matisse French artist (1869–1954)

William Morris British artist and writer (1834–1896)

Pablo Picasso Spanish artist (1881–1973)

palettes specially made piece of wood or plastic for artists to mix their paint on

primary colors red, green, and blue—the colors that mix to make all others

secondary colors the colors made by mixing the primary colors together

shade to add dark areas to a picture

silhouettes when objects are seen against a strong light—they appear as dark shapes

sketchbook a book for making quick sketches to make into paintings later

tapestries a heavy fabric picture or design made by weaving colored threads through it

Victorian the period of history when Victoria was Queen of Great Britain (1837–1901)

Index

Notes for teachers

The projects in this book are aimed at children aged 7-11 years. They can be used as stand-alone lessons or as a part of other areas of study.

While the ideas in the book are offered as inspiration, children should always be encouraged to paint from their own imagination and first-hand observations.

Sourcing ideas

All art projects should tap into children's interests, and be directly relevant to their lives and experiences. Try using stimulating starting points such as found objects, discussions about their family and pets, hobbies, TV programs, or favorite places.

Encourage children to source their own ideas and references, from books, magazines, the Internet, or CD-ROM collections

Digital cameras can be used both to create reference material (pictures of landscapes, people or animals) and also used in tandem with children's finished work (see below).

Other lessons can often be an ideal springboard for an art project—for example, a geography field trip can be used as a source of ideas for a landscape picture.

Encourage children to keep a sketchbook to sketch ideas for future paintings, and to collect other images and objects to help them develop their work.

Give pupils as many first-hand experiences as possible through visits and contact with creative people.

Evaluating work

Arrange for the children to share their work with others, and to compare ideas and methods—this is often very motivating. Encourage them to talk about their work.

Show the children examples of other artists' paintings—how did they tackle the same subject and problems?

Help children to judge the originality and value of their paintings, to appreciate the different qualities in others' work and to value ways of working that are different from their own.

Going further

Look at ways of developing the projects further —for example, adapting the work into collage, print making, or ceramics.

Use image-enhancing computer software and digital scanners to enhance, build up, and juxtapose images.

Show the children how to develop a class art gallery on the school website.